WORLD OF WONDER

Published by Creative Education
123 South Broad Street
Mankato, Minnesota 56001

Creative Education is an imprint of
The Creative Company.

Art direction by Rita Marshall
Design by The Design Lab
Photographs by The Image Finders (Michael Lustbader,
Mark & Sue Werner), JLM Visuals (Breck P. Kent, Zig
Leszczynski, John Minnich), James P. Rowan, Tom Stack &
Associates (Dawn Hire, Victoria Hurst, Breck P. Kent, Thomas
Kitchin, Joe McDonald, Ed Robinson, Tom & Therisa Stack,
Dave Watts), Unicorn Stock Photos (Robert E. Barber,
Richard Gilbert, Russell R. Grundke, Andre Jenny, Thomas
Kitchin, Gary Randall)

Library of Congress Cataloging-in-Publication Data

Hoff, Mary King.
Fighting back / by Mary Hoff.
p. cm. — (World of wonder)
Summary: Discusses how various animals and plants use
special tools or techniques to protect themselves, such as
the turtle's hard shell, the stinging nettles' hairs, and the
pufferfish's ability to expand.
ISBN 1-58341-266-2
1. Animal defenses—Juvenile literature. 2. Animal weapons—
Juvenile literature. [1. Animal defenses. 2. Animal weapons.]
I. Title. II. World of wonder (Mankato, Minn.)

QL759.H45   2003
591.56'6—dc21   2002031488

First Edition

9 8 7 6 5 4 3 2 1

cover & page 1: an underwing moth
page 2: a toxic grasshopper
page 3: a black rhinoceros

Creative Education presents

WORLD OF WONDER

FIGHTING BACK

BY MARY HOFF

Armor-coated insects 🐾 Stinging plants 🌱 Mammals that play dead ❄ The world is full of living things that have special tools or techniques for defending themselves against other creatures that threaten to hurt or eat them.

SOME OF THESE FORMS OF protection are body parts. Others are ways of behaving. Many are a combination of the two. What these defensive traits have in common is that they are **adaptations**. They help improve the creatures' chances of surviving in an eat-and-be-eaten world so they can pass life on to a new generation.

Skunks have one of nature's stinkiest weapons

NATURE NOTE: *A hermit crab protects itself by taking up residence in an empty snail shell. It hauls the shell around wherever it goes.*

TOUGH STUFF

Tasty legs, head, and tail—the slow-moving painted turtle must seem like an easy meal for a **predator** such as a raccoon. But if a raccoon tries to grab it, the turtle draws into its shell. The raccoon is left, empty-pawed, with an object that looks more like a rock than a meal.

Just as knights of old used shields, helmets, and metal suits to protect themselves, many animals use shells and other "tough stuff" to foil their

foes. Clams hide their soft bodies in a hinged shell they open to feed and snap shut when danger approaches. Tortoise beetles and other insects have tough **exoskeletons** that make them less edible to predators. Immature caddisflies, called larvae, live in freshwater streams. They glue together tiny sticks or bits of gravel to make tubes that protect their bodies. Plants use armor, too. Tough coverings such as bark make it harder for insects, rabbits, and other **herbivores** to eat a

Joshua trees are protected by stiff, sharp leaves

plant than if it were as tender as a lettuce leaf. Rice, corn, and other grasses contain a hard substance called silica that some scientists think makes them less appealing to herbivores.

NATURE NOTE: *Silica, a substance found in various grasses, is one of the ingredients used to make glass.*

HORNS AND THORNS

Growths or body parts that inflict injury can be very useful for defense against attackers. Many cactus plants have sharp spines that poke animals that try to eat or otherwise disturb them. So do some palm trees. When a porcupine is threatened, it lashes out with its quill-filled tail. The quills stick into the enemy, causing pain. Large spiders called tarantulas shed tiny barbed hairs from their abdomen that irritate enemies' skin and eyes on contact.

A porcupine's quills can inflict a painful wound

All kinds of body parts can be used for defense. Many animals, including bats, crocodiles, and zebras, use their teeth to fend off foes. Moose use their massive racks to protect themselves from wolves. Kangaroos and ostriches kick attackers with their big feet. Rhinoceroses chase off predators with their intimidating horns.

Some plants use animals as weapons. Acacia trees are protected from herbivores not only by thorns, but also by ants that live in the thorns. These ants attack other animals that try to

Kangaroos fight off enemies with powerful kicks

eat the acacias. Wild tobacco plants being nibbled on by five-spotted hawkmoths release special chemicals into the air. The chemicals attract other insects that attack hawkmoths and their eggs.

NATURE NOTE: *An African elephant tusk can weigh more than 100 pounds (45 kg).*

Tobacco plants rely on bug allies for defense

CHEMICAL POWER

When it comes to defense, chemicals can be a creature's best friend. Of the world's 2,400 or so species of snakes, about 270 produce **venom**, a toxic liquid they inject into other animals' bodies by biting them. Bees and wasps shoot venom into enemies through a stinger on their abdomens. Some kinds of venom are deadly. Others just

NATURE NOTE: *The skin and feathers of the pitohui, a kind of bird found in Papua New Guinea, contain a poison similar to that of South American poison frogs.*

The cobra defends itself with a poisonous bite

cause enough discomfort to discourage future attacks.

☾ The stinging nettle, a plant that grows in moist fields and forests, has tiny, hollow hairs called **trichomes** on its stem and leaves. When a nettle is touched, the hairs break off, releasing chemicals that cause a stinging sensation.

✸ Some plants' chemical weapons cause more subtle harm. Australian eucalyptus trees contain a chemical called tannin that makes it hard for animals to digest their food. Some grasses

and other plants contain chemicals that can interfere with an herbivore's ability to reproduce.

🕷 In the undersea world, scorpionfish ward off foes with their venom-filled spines. Sea anemones, jellyfish, and their relatives

NATURE NOTE: *The platypus, an egg-laying mammal found in Australia, has spurs on its back legs that inject a toxic substance into enemies.*

have stinging parts called **nematocysts** that inject pain-inducing chemicals into attackers. Nudibranchs, or sea slugs, eat sea anemones, then use the anemones' nematocysts to protect themselves from their enemies!

Some creatures produce chemicals that sicken or kill animals that try to eat them. Certain mushrooms can kill an animal when eaten even in tiny quantities. Many Central and South American frogs have poisonous substances on their skin that can be deadly to predators.

Most sea creatures avoid the anemone's sting

Some animals that use chemicals to defend themselves warn potential attackers first. Scorpions, snakes, and other creatures make rattling or hissing sounds. Poisonous frogs, bad-tasting insects, and other animals protected by chemicals tend to be brightly colored. Known as **aposematic coloration**, this serves as a warning to foes to stay away.

NATURE NOTE: *Some plants produce chemicals that, when eaten by insect larvae, prevent them from turning into adult insects.*

Rattlesnakes will give warning before striking

SURPRISE AND DISGUST

Skunks are well-known in North and Central America for the nasty-smelling liquid they spray when threatened. Many other kinds of living things also ward off enemies using substances or behaviors that surprise or disgust.

❂ Underwing moths escape predators by moving their top wings to reveal brightly colored wings beneath. The bright colors startle the attacker, giving the moth time to escape. Bombardier beetles spray a jet

NATURE NOTE: *A skunk can hit an enemy with its stinky spray from up to 12 feet (4 m) away.*

A skunk's foul spray is difficult to wash off

of hot liquid at animals that get too close. Horned lizards, found in North America, squirt blood from holes near their eyes. Fulmars, a type of ocean-going bird, scare predators from their nests by vomiting a repulsive liquid at them.

Milkweeds have sticky, bad-tasting sap that deters most insects from eating them. Monarch butterflies take

NATURE NOTE: *The pangolin, a scale-covered mammal found in Asia and Africa, rolls into a ball when threatened. It can also spray a bad odor to frighten enemies away.*

The monarch butterfly is a bitter-tasting meal

advantage of the milkweed's defense system. By feeding on the sap, they become bad-tasting, too. A bird that eats a monarch quickly learns to avoid such butterflies.

NATURE NOTE: *Boxer crabs protect themselves by carrying stinging sea anemones around with them.*

SAFETY IN NUMBERS

Many animals use the strategy of mingling with others of their kind as a way to reduce their odds of getting eaten. Penguins build their nests in large groups called rookeries.

NATURE NOTE: *Musk oxen are large, hairy animals that live in the Arctic. When they are threatened, they line up or form a circle with their horns pointing toward the enemy.*

Herring swim together in the northern oceans by the millions. African wildebeests, Arctic caribou, Arabian oryx, and many other hoofed animals travel in herds.

❀ Hanging out in groups also lets animals warn each other of danger. When some ants, bees, and other insects are frightened, they give off airborne chemicals called **pheromones** that alert

NATURE NOTE: *When an enemy approaches a school of mullets, the fish split up and swim in two directions, confusing the foe.*

Caribou move in herds on the Arctic tundra

others of their kind that trouble is near. Some fish give off a chemical when they are injured that frightens their companions away. Some birds use strength in numbers to chase enemies away from their nests. Birds such as chickadees and crows use alarm calls to round up other birds, then group together to attack a hawk or other enemy that is threatening them.

NATURE NOTE: *Some animals zig-zag when they run away. This helps them by making it harder for the enemy to predict which way they will go.*

A school of fish may number in the thousands

RUN AND HIDE

Legs, fins, and wings can provide powerful protection by allowing an animal to move quickly to escape danger. Squid do even better. Before they propel themselves away from an enemy, they squirt a black, inklike liquid from their body to distract the foe.

☀ Some animals protect themselves by hiding from their enemies. In the winter, meadow voles burrow beneath the snow, where they aren't as obvious to hungry hawks and owls as they would be on the surface. The shield-tailed

25 *A squid uses "ink" in a unique disappearing act*

snake, found in Asia, dives into a hole when threatened, then blocks the opening with its stubby tail.

◆ Plants can't run, but some can hide! Some mistletoe plants grow in the foliage of Australian eucalyptus trees. Because they have leaves that look a lot like the leaves of the eucalyptus, the mistletoe plants can elude animals that may want to eat them.

⚜ Some lizards, sea slugs, centipedes, spiders, and other animals add an interesting

This skink (a lizard) has lost the tip of its tail

twist to their escape routine. They leave a body part such as a tail or limb behind when they flee. This is called **autotomy**. The predator is distracted by the left-behind part, which may still be moving, while the animal escapes.

NATURE NOTE: *Flying fish escape predators by launching themselves from the water into the air. Some flying fish can soar as far as 1,000 feet (300 m).*

TRICKY LOOKS

Some creatures protect themselves by fooling predators into thinking they are something dangerous. The larva of a certain kind of Costa Rican moth has markings that make it look like the head of a deadly snake. Some species of treehoppers (a type of insect found mainly in the tropics) look like thorns.

✳ Animals can also ward off enemies by making themselves look big and tough. Pufferfish expand by filling themselves with water or air when threat-

The pufferfish can transform into a spiky ball

ened. Cobras and some other snakes raise their heads and widen their bodies to appear more ominous. A North American bird called the ruffed grouse defends its nest by rushing at a predator with its feathers fluffed and wings outspread.

✳ The killdeer, a bird that nests on the ground, protects its young with another kind of trick: it pretends it's injured. When a predator approaches, the killdeer leaves its nest dragging a wing as though it were broken. The predator follows the parent bird, which then escapes. Other birds that nest on the ground, including piping plovers, short-eared owls, and the dotterel (a New Zealand bird), also use the "broken-wing" act to prevent their nests from being raided.

🐾 Opossums play an even better trick than pretending they're wounded. They pretend they're dead! Hognose snakes also use this ploy to keep from being eaten.

NATURE NOTE: *Hover flies scare off predators by mimicking stinging insects. Some even imitate the buzz of a bee!*

AMAZING ADAPTATIONS

From deserts to bogs, from mountains to oceans, the world is filled with plants and animals that protect themselves against enemies in a variety of ways. Some use common and obvious defenses, such as spines or teeth. Others use more subtle approaches, such as hiding or playing dead.

These defenses are just a few of the countless adaptations that help living things thrive in the challenging world around them. They are valuable reminders of how intricately the lives of various creatures are intertwined. As humans make changes that affect the environment, it's important to remember and respect these connections. In doing so, we can help ensure the future health and beauty of this amazing world, this world of wonder.

NATURE NOTE: *The window pane palm grows leaves that look like they've been chewed on. Scientists think this makes them less desirable to herbivores.*

The horned lizard squirts blood at its attackers

WORDS TO KNOW

Adaptations *are characteristics that contribute to a living thing's ability to survive or reproduce.*

Aposematic coloration *makes a creature stand out; it sends a message to enemies that the creature is dangerous.*

The act of shedding a body part when confronted by an enemy is called **autotomy**.

Insects are protected by hard outer coverings called **exoskeletons**.

Animals that eat plants are called **herbivores**.

Nematocysts *are stinging body parts found on jellyfish and their relatives.*

Chemicals that living things use to communicate with each other are called **pheromones**.

A **predator** *is an animal that kills and eats other animals.*

Trichomes *are tiny hairs found on some kinds of plants.*

Venom *is a substance produced by snakes, bees, hornets, and other animals that causes pain or death when injected into another animal.*

INDEX